PROTEINS

Rebecca Woodbury, Ph.D., M.Ed.

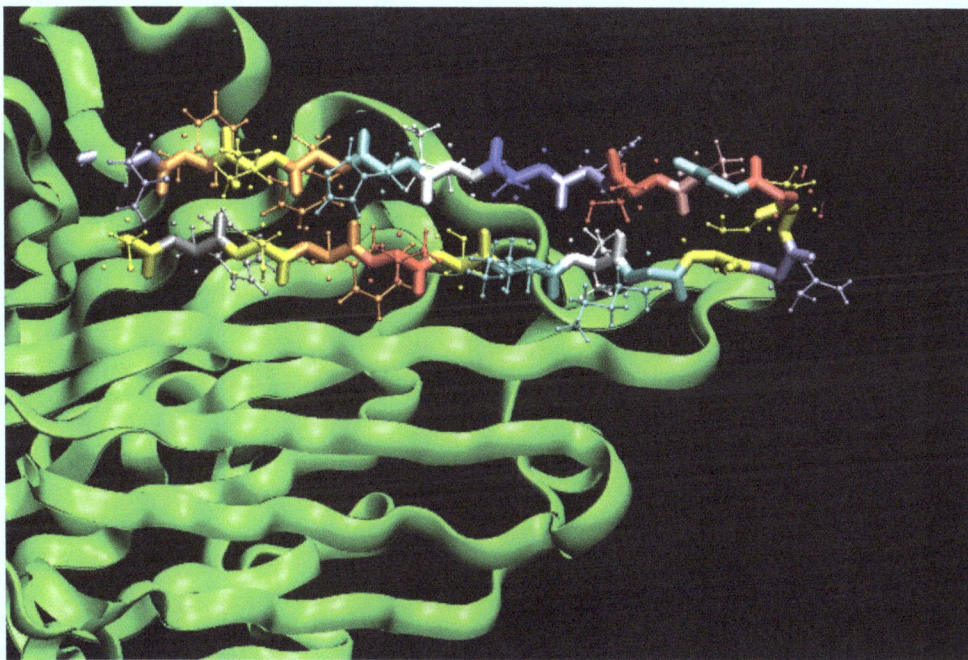

Gravitas Publications Inc.

PROTEINS

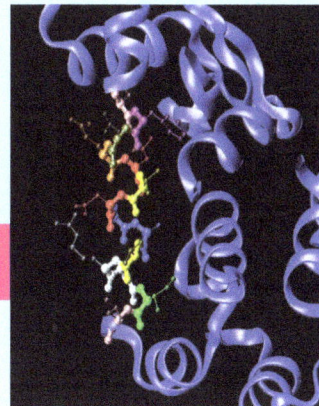

Illustrations: Janet Moneymaker

Copyright © 2024 by Rebecca Woodbury, Ph.D., M.Ed.

Proteins
ISBN 978-1-950415-16-8

Published by Gravitas Publications Inc.
Imprint: Real Science-4-Kids
www.gravitaspublications.com
www.realscience4kids.com

RS4K

Image credits: Cover, Title page, Above, & Page 15: Crystal structure by Habash, J., Raftery, J., Nuttall, R., Price, H. J., Wilkinson, C., Kalb, A. J., Helliwell, J. R. "Direct determination of the positions of the deuterium atoms of the bound water in concavalin A by neutron Laue crystallography" *Acta Crystallograhphy*, Section D v56 pp. 541–550, 2000 (Protein Data Bank ID 1C57). Illustration by D. J. Keller; Page 7. Meat, By New Africa, AdobeStock; Milk, Image by Couleur from Pixabay; Beans, Image by pixel1 from Pixabay; Cheese, Image by Ralf from Pixabay; Page 21, By Robert Kneschke, AdobeStock

Have you ever wondered

what a protein is?

I don't know.

Me either.

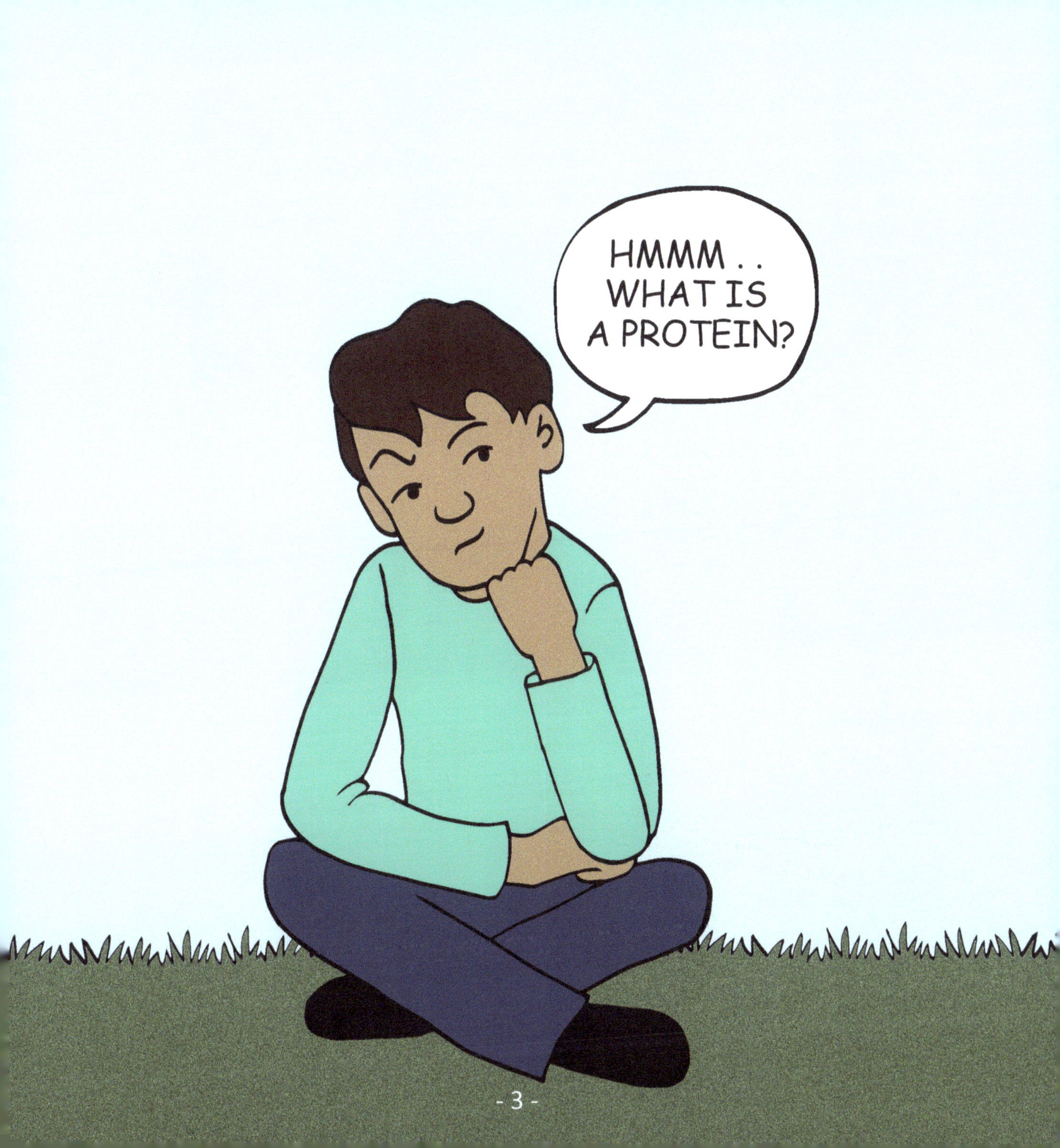
HMMM . .
WHAT IS
A PROTEIN?

You may have heard someone say,

"Eat more protein!"

Eat your cheese!

No problem.

Maybe you know that
meat, milk, cheese, and
beans have protein.

Cheese has protein?

Yes!

But what is a protein?

Protein?

Protein?
Turn the page
to find out.

Proteins are special **polymers.**

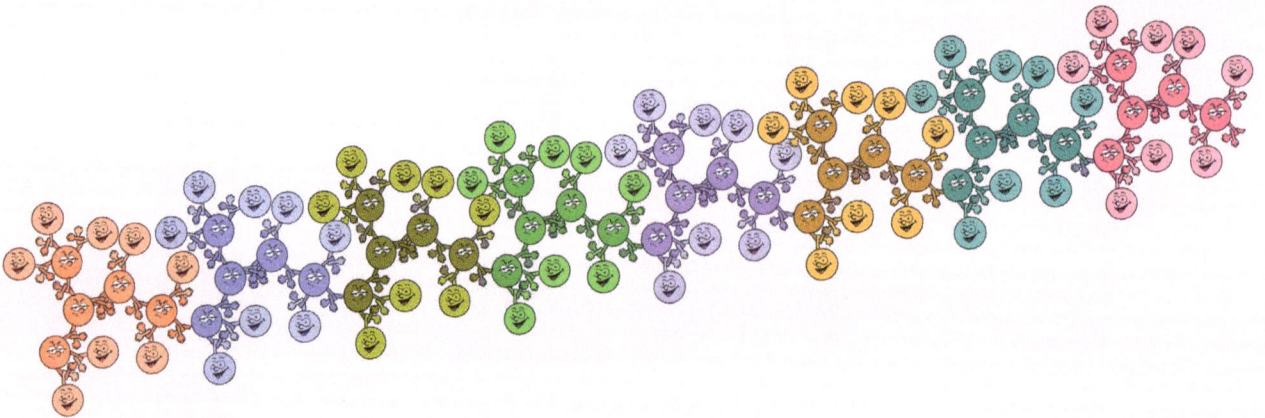

Review: POLYMER

A **polymer** is a **molecule.**

It is a long chain of atoms linked together.

"Poly" means "many."

"Mer" means "unit."

The word **polymer** means **many units.**

Review: ATOMS

- **Atoms** are tiny building blocks that can link together.

- **Atoms** make everything we see, touch, taste, and smell.

Review: MOLECULES

Molecules are made
when **atoms link** together.

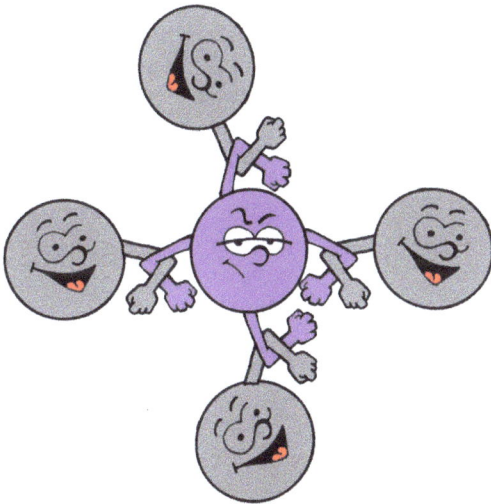

Protein polymers fold

into different shapes.

This is a protein polymer folding into a coil shape.

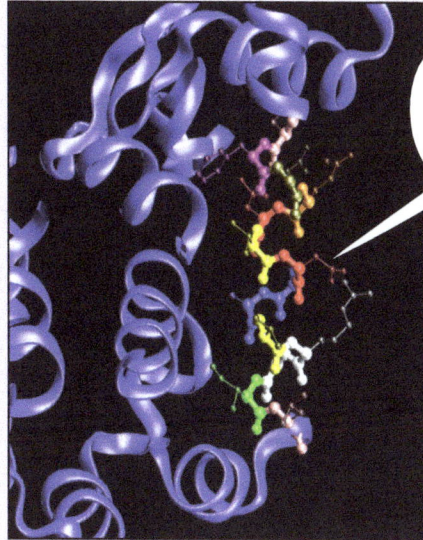

We are a chain of atoms in the protein molecule.

This is a protein polymer folding into a sheet shape.

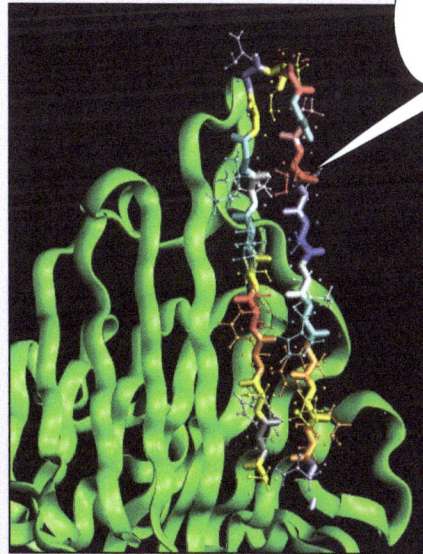

We are a chain of atoms in the protein molecule.

These different shapes
allow proteins to behave
like tiny machines.

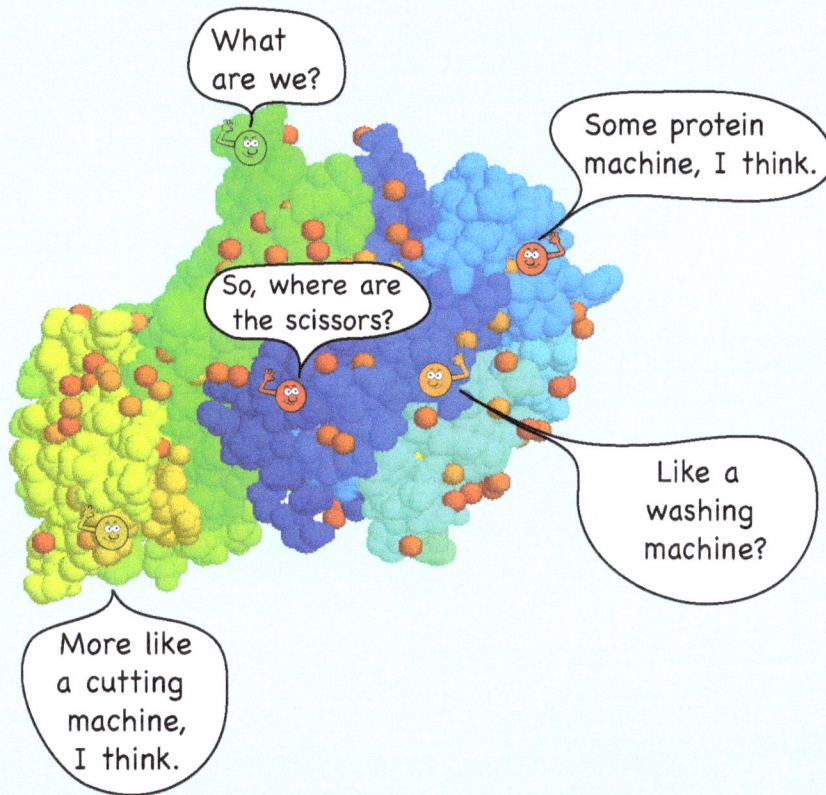

Protein machines do all the
big jobs inside your body.

Some protein machines
glue molecules
together.

Some protein machines
read molecules.

Some protein machines
cut molecules.

Some protein machines
move molecules.

By cutting, gluing, moving, and reading molecules, proteins help your body grow big and strong!

That is why I eat cheese!

How to say science words

atom   (AA-tum)

machine   (muh-SHEEN)

molecule   (MAH-lih-kyool)

polymer   (PAH-luh-muhr)

protein   (PROH-teen)

science   (SIY-uhns)

www.ingramcontent.com/pod-product-compliance
Lightning Source LLC
Chambersburg PA
CBHW040151200326
41520CB00028B/7571